THE REAL
STORY

CAMBODIA AND SOUTH VIETNAM 1953–1970

TERENCE MCCARTHY
USA (LTC) Retired

Copyright © 2022 by Terence McCarthy.

All rights reserved. No part of this publication may be reproduced, distributed, or transmitted in any form or by any means, including photocopying, recording, or other electronic or mechanical methods, without the prior written permission of the author, except in the case of brief quotations embodied in critical reviews and certain other noncommercial uses permitted by copyright law.

Printed in the United States of America.

ISBN	Paperback	978-1-68536-591-2
	eBook	978-1-68536-592-9

Westwood Books Publishing LLC
Atlanta Financial Center
3343 Peachtree Rd NE Ste 145-725
Atlanta, GA 30326

www.westwoodbookspublishing.com

TABLE OF CONTENTS

Introduction 1
The Real Story 5
Preface. 9
Prologue: The Vietnam Memorial Poem 10
Addendum 12
Cambodia/South Vietnam 1953-1970 16
President's Foreign Intelligence Advisory Board
Letter 88

INTRODUCTION

By
Frank T. McCarthy (LTC RETIRED)
Edited By
Matthew Cory

The author, Major F. Terence McCarthy, a US Army Intelligence officer, was assigned to U.S. Military Assistance Command, Viet Nam (COMUS MACV) in August 1968 as the chief, Out Country Section in the Intelligence Production Division.

He examined all of the available intelligence data concerning the Cambodian situation and developed some new intelligence estimates, which clearly indicated that the Cambodians were not only allowing arms shipments destined for the Vietnamese communists to be off-loaded at the port of Sihanoukville, but were also involved in the transport of munitions across the country to Vietnamese/Cambodian communist sanctuaries located along the Vietnamese-Cambodian border. These estimates, dispatched to Washington, were not met with favorable nods of agreement there. The Washington community still clung to the belief that the Ho Chi Minh Trail provided the bulk of military supplies to communist forces in the south part of South VietNam. They claimed that "hundreds of bits and pieces of information" supported this view.

Major McCarthy, aware of this reasoning, devised a strategy to resolve this major dichotomy. He set out to prove or disprove the Ho Chi Minh Trail theory. The use of Cambodian border regions as Vietnamese base camps was not in dispute. The Washington policymakers shrugged that situation off saying that the Cambodian government's control was very weak. The new strategy was the development of a new Cambodian collection plan called **BLACKBEARD**.

This new approach discarded the traditional, blank, encyclopedic information-gathering methodology. Instead, **BLACKBEARD** provided collection units with basic intelligence holdings in the targeted area and asked them to verify these holdings. Top priority was given to reporting on any military supply flow southward through Cambodia's eastern provinces of Ratanakiri and Mondolkiri. The only continued communist activity, which was detected in these areas, was the movement of personnel replacement packets.

His second ploy was a request that he be sent to MACV's Combined Intelligence Center, Vietnam (CICV). In that request, he asked CICV to estimate the Vietnamese communist monthly expenditure of ordnance in southern South Vietnam over the last year, to reduce that figure by any known capture of usable arms and ammunition from friendly forces, and to build a logistics model on the net munitions remaining (see attached). This would depict the size and level of activity of Vietnamese communists support forces, which would be required to be in place in Eastern Cambodia if, indeed, the Ho Chi Minh Trail was involved. The resultant model described a logistical network of awesome proportions. All-source intelligence sources including many agents traversing the possible overland routes did not detect any evidence of military supply flow. The Washington community's "hundreds of bits and pieces" of support for the Ho Chi Minh Trail was literally blown out of the water.

Major McCarthy was subsequently transferred to the US Army Staff at the Pentagon in the fall of 1969. He continued to provide significant input on several Indochina projects to include the writing of a history of the Cambodian involvement in the Viet Nam conflict. In 1970, he appeared before the secretary of the President's Foreign Intelligence Advisory Board (PFIAB) and reviewed with him the overwhelming nature of the evidence concerning Cambodia and the MACV estimates, which were cabled to the Chairman of the Joint Chiefs of Staff. Subsequent to this meeting, it was rumored that the President favorably acted upon the PFIAB's recommendation that the CIA be censured for its role in keeping the differing MACV Cambodian estimates from being noted in National Intelligence estimates previously submitted to the Nixon White House.

The Vietnamese Communist Use of Cambodia:

THE REAL STORY

By
Frank T. McCarthy (LTC)-Retired

Over the years the media has portrayed very inaccurately the major role Cambodia played in the Vietnam War.

What is still very distressing is the "Real Story" has never been told of the failure of the United States to properly act on the story of the actual use of Cambodia in, which I was personally responsible for achieving an intelligence break-through in the Fall of 1968 in Saigon. Had the United States reacted in a responsible manner on this intelligence, I believe that many American servicemen would not have been killed and many more would not have experienced serious wounds that they received from the Vietnamese Communists who were using Cambodia base areas to launch offensive operations. After initiating assaults in South Vietnam, many communists units retreated to safe havens in Cambodia. Our policy was that we could only engage them IF we were in pursuit of them as they crossed the border. For the most part, the Communists who knew what the policy was made significant efforts to ensure that pursuit wasn't happening when they crossed the Cambodia/South Vietnam border.

The most glaring inaccuracy in media reporting of the Vietnam War deals with the flow of logistics and ammunition to the Vietnamese Communists in Southern South Vietnam. In the fall of 1968 through the compilation of intelligences from all sources and the piecing together a comprehensive estimate of the operations in the Cambodia/South Vietnam region, I was able to prove beyond a reasonable doubt that no significant flow of ordinance/ammunition was reaching Vietnamese Communist forces in Southern Vietnam by sea or over the Ho Chi Minh Trail. Instead, I conclusively demonstrated that ordinance/ammunition shipments were delivered to the port of Sihanoukville, moved to a transshipment depot, Kompong Speu, southwest of Phnom Penh and was further transported to Vietnamese Communist units in the Cambodia base areas along the South Vietnam border.

I briefed these findings to General Creighton Abrams (the US Vietnam commander) in October 1968 and he had me write back channels containing these findings to the Joint Chiefs of Staff and the Pentagon on several occasions.

General Abrams was hoping that the Vietnamese Communist use of Cambodian sanctuaries could be eliminated - saving many American lives. And that some punitive action could be taken to prevent the flow of ordnance into and from Sihanoukville.

No action from Washington was forthcoming except to send a high level intelligence team out to Saigon in December '68 to put an end to my estimates and findings. They were completely unsuccessful in this effort.

General Abrams was very upset with all of this foot dragging. Over Christmas in 1968, Billy Graham visited the troops and stopped by Saigon. In an effort to get this critical estimate on Cambodia to President Nixon, General Abrams invited me to his quarters and had me give Mr. Graham a complete briefing on the Cambodian situation.

Over the ensuing months of 1969, our case on Cambodia became even stronger.

In July of 1969 General Abrams had me pack up my briefing and sent me to Washington to brief the entire community - which I did.

After my tour in Saigon was over, I was assigned to the Pentagon and became General Westmoreland's chief go-to-guy on Cambodia and Vietnam.

Very briefly, I put together a study on Cambodia, which I briefed to the President's Foreign Intelligence Advisory Board. They brought the matter to the attention of the President, who censured the Director of Central Intelligence for failing to convey the Vietnam Commander's estimates to his office.

A couple of years ago, through the FOI Act I was able to get the Pentagon study declassified and released.

Frank T. McCarthy, LTC, USA-MI(RET)

PREFACE

As a member of the US Army, I was stationed in South Vietnam in 1968 and 1969. Many of my friends were killed or seriously wounded in that conflict. My reception, when I returned home to Cedar Rapids, Iowa, in 1969 was far from friendly. Over the years, I had suppressed my hurts and frustrations generated by the Vietnam experience. So, it was with great surprise, and relief, that my initial visit to the Vietnam Memorial in Washington D.C., resulted in a torrent of tears and an eruption of emotional revulsion. The catharsis that the Vietnam Memorial provided, was ironically refreshing. This healing response continues to be even more strongly felt each time I return to this silent, yet awesome, tribute to those who gave so much for such a losing cause.

PROLOGUE
THE VIETNAM MEMORIAL

The Wall

The black, marbled wall was so greatly imposing,
Standing there naked with over 58,000 names,
Which have finally found a dignified reposing
Near Lincoln, Jefferson and JFK's memorial flames.

How dumb struck and gushing with purgative tears
Was I on that misty Sunday morning in eighty three;
Engulfed in the revulsion of defeat and fallen peers,
I felt the profound Vietnam burdens going on a spree.

How extraordinary to give up such grisly baggage,
Not knowing one white of its implacable, deadly load.
The wall had freed me from an Asian bondage,
Giving me courage to look ahead along life's long road.

Even though one thought lingers: All have not died in vain;
The Vietnam Memorial moves one to a level quite humane.

*Please note: The author of this poem is Terence McCarthy. This poem is included in the 50 rhymed Shakespearean sonnets that make up his published book that is entitled <u>Poetry with Feeling</u>. The book is available at Amazon.com.

ADDENDUM

The author is providing some added supporting documentation to reinforce the factual commentary in this study.

Major McCarthy arrived in Saigon in August, 1968.

<u>Assignment: J-2 MACV</u>, Chief of Out Country Section.

In his first 40 days analyzing the Out Country situation, with special emphasis on Cambodia, he established the following contacts:

1. American Embassy twice weekly – he drove his own Jeep from Tom Som Nhut Airport and visited the following;
   a. US Embassy, Cambodian analyst;
   b. CIA station chief

2. US Naval Forces, Vietnam (NAVFORV) once weekly;
3. 5th Special Forces - twice weekly;
4. US Army MACV SCIF (daily) to include NSA analysts and nationwide communist activities;
5. Monthly phone conversations with Australian attaches of the Phnom Penh Australian Embassy. They were helpful in providing current data on Prince Sihanouk.

After 40 days, Major McCarthy assembled an estimate which included the following:

    a. Cambodian meeting CHICOM cargo ships, which were delivering arms and ammunition destined for Cambodian sanctuaries and Vietnamese troops.

    b. Cambodian Army (FARK) moving the armament supplies to a depot south of Phnom Penh and onward to communist forces.

    c. Redefining sanctuaries located along the Cambodian/South Vietnam border.

    d. The US Army 5th Special Forces established several base camps located near eastern Cambodia's Mondolkiri and Ratanakiri Provinces. The base camps included several Montagnard units which served to surveil the trails used by the NVA to move personnel packets to southern South Vietnam. No movement of arms and ammunition were detected. The trail was only used by foot traffic since no roadways existed. This constant surveillance proved that the Ho Chi Minh Trail did not extend south of the tri-border area.

e. USNAVFORV was contacted once a week. It ran the Market Time Coastal Operations, which completely denied the NVA the delivery of supplies to the total seacoast of the RVN after mid-1965.

f. The NSA element at MACV SCIF provided complete coverage of communications of VC/NVA units throughout the South Vietnam theater. They were contacted several times every week to identify VC/NVA movements in Cambodia and southern South Vietnam.

g. B-52 strikes were scheduled each Tuesday afternoon at 4:30 p.m. in the MACV SCIF. The NSA intercepts were a major factor in targeting. Even though Cambodian sanctuaries were off limits, those VC/NVA just entering RVN were a priority.

CAMBODIA/SOUTH VIETNAM
1953–1970

By
FRANK TERRENCE McCARTHY
USA (LTC) RETIRED

10/1968

SUMMARY: 1964-1970.

A. POLITICAL.

1. From November 1953 when Cambodia became independent until March 1970 when Sihanouk was overthrown, the primary concern of Sihanouk's foreign policy was to preserve the independence and territorial integrity of Cambodia. In his view, the chief threats to these objectives were posed by Thailand and Vietnam. For several centuries these countries had encroached on Khmer territory, and Sihanouk feared that they still had designs on Cambodia.

2. At the time of gaining independence, Cambodia was small, weak, and in a vulnerable position. Consequently, to fulfill his foreign policy objective of preserving Cambodia's independence and territorial integrity, Sihanouk was faced with the decision of joining one of the two great power blocs or of assuming a neutral position. By late 1954, he decided on the latter course, proclaiming in December that Cambodia would remain nonaligned and would adhere to an official policy of neutrality. This was a conscious effort to balance Western influence with Communist ties. It was also partly a reaction to the influence of Nehru and India and to the growing vogue of neutralism in the mid-1950's.

3. For the next decade, Sihanouk held to this neutralist course. In September 1957, he had Cambodia's National Assembly pass the Neutrality Act, which declared that Cambodia would abstain from aggression, while reserving the right to self-defense. Three years later, Sihanouk proposed to the United Nations that a neutral ZONE composed of Laos and Cambodia be established and guaranteed by the MAJOR power blocs. In both 1962 and 1963, he proposed a MULTINATIONAL moratorium

UNCLASSIFIED

- 17 -

conference to guarantee Cambodia's independence, neutrality and territorial integrity.

4. In late 1963, although still working within the framework of an official foreign policy of neutrality, Sihanouk began his drift to the left, as evidenced by his termination of US aid in November. The reasons for this shift rested in Sihanouk's reappraisal of the US role in Southeast Asia, his belief of an emerging hegemony of Communist China, and his fears of an intensified Vietnam War, which he believed North Vietnam would eventually win. His primary concern remained the survival of Cambodia, and he believed that a movement toward the Communist Bloc would be the best guarantee of Cambodia's survival. Cambodia's swing to the left continued until Sihanouk's ouster, although there were periodic retrogressions during times when Sihanouk desired to keep some semblance of neutrality for the purpose of appeasing growing Western apprehensions.

5. During 1965, Sihanouk turned abruptly from the United States and on 3 May announced that he was severing diplomatic relations with the US. It was noted in October 1965 that Cambodia, having departed from its enunciated policy of neutralism, had become aligned in a sense with the Communist Bloc. Sihanouk was making efforts to ingratiate himself and his nation with the leaders of "the wave of the future" in order to preserve his nation. One of the steps taken by Sihanouk was to accept a Chicom military aid program and to enter into a Chinese-Cambodian cultural agreement. In his public addresses in 1965, Sihanouk emphasized Cambodia's friendship with China, which, he said, had promised to stand behind his country in its struggles in the avenue.

6. This trend continued in 1966, and MACV J2 noted a growing Chicom influence in Cambodia. Military intelligence in Saigon felt that Sihanouk's motives had not changed and that he was orienting his country toward Red China as a result of his belief that it was the best way to safeguard Cambodia's sovereignty and territorial integrity with himself in power. Sihanouk remained chary in his relations with South Vietnam and Thailand, using the occasion of periodic border incidents to warn of the expansionistic tendencies of these neighbors. [■■]

Despite his close association with the Communist Bloc, Sihanouk argued that his country was still neutral. He offered on several occasions the proposal that the International Control Commission (ICC) be expanded so that it could effectively patrol the border. The Soviets and Chinese objected to the proposal, however.

7. The first major retrenchment in Cambodia's close association with China occurred in 1967. Foreign military observers in Phnom Penh noted that Sihanouk was becoming concerned both with China's growing influence in Cambodia and with the militant activities of the domestic Cambodian Communist rebels, the Khmer Rouge. In order to keep a balance between the Western and Eastern influences in his country, Sihanouk verbally attacked the Chinese Communists on several occasions, especially after the Khmer Rouge-inspired rebellion in Battambang Province in early 1967. At one point, he threatened to close his embassy in Peking. However, after the receipt of a conciliatory letter from

- 19 -

Premier Chou En-lai in September, Sihanouk softened his hard stand against Peking. During the year, however, he persisted in his chilly attitude toward the US and the Allies. He adamantly denied US claims that the enemy was using Cambodian territory, arguing that these claims indicated the Allies' intentions to expand the war into Cambodia. However, to give evidence of his neutral posture, Sihanouk again extended offers to expand the role of the ICC to guarantee the inviolability of Cambodian soil. Also, he allowed American newsmen to tour Cambodia without much restriction. However, following the discovery and report by newsmen in November 1967 of a Vietnamese Communist base camp in the Himot area, Sihanouk quickly denied the story and refused future examination of the border area by Western journalists.

8. There was a temporary swing away from further rapprochement with the Vietnamese Communists in 1968, for Sihanouk was becoming aware of the extent of their use of Cambodian territory. The role of Cambodia as a supply base for the enemy in the Tet Offensive tended to exacerbate the Prince's quest for guaranteed neutrality for Cambodia. Sihanouk was said to be particularly upset and enraged when he saw the US documentation of Viet Cong activities in Cambodia. This raised hopes of a closer alignment for Cambodia with the United States. Nevertheless, through the course of the year, Sihanouk kept denouncing Americans for not wanting peace in Vietnam, and he leveled acerbic attacks against the US as a result of border incidents. Relations with Communist China remained unchanged in 1968, although Sihanouk did express concern over possible Chinese denunciation of Cambodia in the future. The Prince expressed hopes that upon leaving SVN, the US would remain in Thailand and the Philippines.

UNCLASSIFIED

- 20 -

to offset Chinese influence.

9. During 1969, Prince Sihanouk's foreign policy tactics included vacillating between the Vietnamese Communists and the United States. Three primary factors influenced his political maneuvering: (a) Sihanouk realized that the VC/NVA were not winning the war and that the struggle had become much longer than originally anticipated; (b) Allied military success in South Vietnam had caused the development of large, permanently garrisoned VC/NVA enclaves in Cambodia, over which Cambodia had limited or no sovereignty; and (c) Sihanouk feared the indigenous Communist insurgency in Cambodia and the aid which the VC, NVN and Communist China were giving it.

10. As a result of these factors, Cambodia drew closer to the US in 1969. In March, Sihanouk admitted that the VC/NVA were infiltrating through Cambodia and occupying parts of his country. Later he spoke for the first time of possibly severing relations with Hanoi. As relations with the US improved in the spring, Sihanouk moved to cut off the flow of munitions and other supplies to the VC/NVA by placing an embargo on the transshipment of arms and supplies from the depots at Kompong Speu and Lovek. This embargo lasted until mid-September 1969, when Sihanouk apparently   understanding with the leaders of the Viet Cong

and North Vietnamese during a meeting in Hanoi.

11.

The principal means of Allied pressure for a change in Cambodian foreign policy was the threat of hot-pursuit or cross-border operations directed against VC/NVA installations in the border regions. Moreover, the possibility of improved US relations carried with it the prospect of significant aid assistance, which would help to reduce Cambodia's economic strains.

12. The severe economic troubles caused Sihanouk to install Lon Nol as Prime Minister on 12 August 1969, charging him with the responsibility of rectifying the economic problems. Since Lon Nol was almost continually absent because of personal reasons and state visits, Deputy Prime Minister Prince Sisowath Sirik Matak, for all practical purposes, served as Prime Minister. Matak's relations with Sihanouk deteriorated during the period over the question of how to implement necessary economic reforms. Moreover, it was reported that between August and the end of 1969, Sihanouk made three attempts to dissolve the government through political maneuvering. However, he met opposition--the first time in a decade that he had faced concerted resistance to his policies.

13. With both Sihanouk and Lon Nol out of Cambodia in early

Khmer

1970, Matak was in charge of the government for about a month and a half.

When Lon Nol returned to Cambodia on 18 February, the Royal Khmer Government expressed new concern about several fundamental problems with the Vietnamese Communists. On 11 March, the North Vietnamese and the Provisional Revolutionary Government (PRG) embassies were sacked by thousands of students. These assaults were preceded by anti-VC/NVA demonstrations in several provinces. The attacks received the unanimous support of the Cambodian legislature, which passed a declaration requesting the government to take all measures necessary to solve the problem of enemy infiltration. Finally, on 18 March 1970, the National Assembly unanimously voted Sihanouk out of office as Head of State. Subsequently Sihanouk was stripped of his military and political party titles, and the army was ordered "to crush by means of arms all actions which Prince Norodom Sihanouk may provoke."

14. Seeking refuge in Peking, Sihanouk announced in a series of radio broadcasts from China his intention to lead the struggle to oust the Lon Nol-Sirik Matak government. He announced the formation of an exile government--the Royal Government of National Union. He also told of plans to organize the newly-formed National United Front of Kampuchea at the grassroots level throughout Cambodia. The North Vietnamese and Viet Cong, faced with a threat to their Cambodian sanctuaries and resupply haven, moved with great speed following the 18 March dismissal of Sihanouk. They verbally supported Sihanouk, harshly denounced the new regime, withdrew their embassy personnel, suspended diplomatic relations, and

- 23 -

tried to create and exploit civil chaos. The Chinese Communists applied pressure on the Lon Nol government by providing Sihanouk with a forum and by offering aid for the liberation of Cambodia. Meanwhile, the Soviet Union adopted a posture of extreme caution toward events in Cambodia.

15. Saigon lost little time in capitalizing militarily on Sihanouk's ouster. In early April, ARVN troops began the first of a series of forays across the Cambodian border into Communist base areas. Concurrently, the Communists began stepping up their attacks on major towns and lines of communication in areas south and east of Phnom Penh. The subsequent massive Allied cross-border attacks against enemy military sanctuaries in May and June tended to put an official and long-term stamp on the emerging military relationship between Phnom Penh and Saigon, especially in view of the fact that continuous (ARVN) support would be Army of Republic of Vietnam necessary until the Cambodian Army could assume the bulk of the combat burden.

16. With Sihanouk's ouster, the most important cause of discord in Thai-Cambodian relations was removed. The meeting between Foreign Ministers Thanat and Yem Sambaur in Bangkok on 3 May resulted in agreement on the following points: (a) South Vietnamese troops were operating in Cambodia with the agreement of the Government of Cambodia (GOC); (b) diplomatic relations were to be established; (c) the GOC agreed to protect the lives and goods of Vietnamese residing in Cambodia and to take all necessary measures to protect the security of Vietnamese who desired to stay in Cambodia; and (d) each country pledged itself to respect the border of the other.

UNCLASSIFIED

MAP

17. As the year ended, a 255-million-dollar MAP authorization for Cambodia was approved by Congress on 22 December. It contains the proviso that no US ground combat troops could be introduced into Cambodia, unless the President deemed such a step necessary to promote the safe and orderly withdrawal of American troops from South Vietnam or to obtain the release of Americans held as prisoners of war.

UNCLASSIFIED

B.

a. VC/NVA USE OF CAMBODIA.

1. The Vietnamese Communists' use of Cambodian-territory over the years was poignantly portrayed in the following comments:

b. On 24 February 1968 at a news conference, General Westmoreland said:

"The enemy has ignored the neutrality of Cambodia . . . and has taken advantage of this situation and has used the border areas of Cambodia for purposes of infiltration, supply and troop rehabilitation and training. Even in the populated areas of Cambodia adjacent to South Vietnam, a smuggling operation of major proportions has been conducted to supply the Viet Cong forces with arms, ammunition and medical supplies."

c. Then, on 30 June 1970 in commenting about the Allied cross-border operations in Cambodia, President Nixon stated:

"The prospect suddenly loomed of Cambodia's becoming virtually one large base area for attack anywhere in South Vietnam along the 600 miles of the Cambodian frontier. . . . The possibility of a grave threat to our troops in South Vietnam was rapidly becoming an actuality."

2. Even though little information was available in 1964 concerning Vietnamese Communists installations and supply centers on

Cambodian soil, MACV estimated that the Viet Cong were probably using Cambodian territory for recuperation, regroupment and resupply. Moreover, the importance of Cambodia to the enemy was underscored in a document taken from a VC prisoner in December 1963 who headed the Psychological Warfare Section, Central Guerrilla Headquarters. It stated that the neutrality of Cambodia and Laos facilitates the expansion of secure base areas, without which "we cannot develop our mission." Five areas along the Cambodian-Republic of Vietnam border were suspected of being Viet Cong sanctuaries.

3. By the end of 1965,  reported that the evidence showing the Vietnamese Communists' reliance on Cambodian territory had become indisputably conclusive. This evidence revealed that the Communists used their Cambodian sanctuary regions for the following reasons:

       a. As a refuge when pursued by American or ARVN forces.

       b. As a training or hospital area.

       c. As a staging area.

       d. As a logistical area.

       e. As safe infiltration or transit routes.

       f. As a communications zone.

4. The importance of Cambodia to Communist forces for supporting their offensive actions in South Vietnam became abundantly clear in October 1965 when three North Vietnamese Army (NVA) Regiments staged out of Cambodia for the Ia Drang Valley battle with US forces. Several of the more than 100 prisoners captured in this battle told of their passage

through and training in Cambodia prior to the engagement and the movement of supplies from Cambodia to the battle area. Moreover, intelligence indicated that small groups of Communist forces had taken refuge in Cambodia following the Ia Drang Campaign.

5. In a study in September 1966, MACV J2 stated that, since the latter part of 1965, there had been a marked increase in reported identifications of NVA and VC units in Cambodia. This trend was attributed to the increased and practically unrestricted use of Cambodia by units that moved across the border for reorganization, rehabilitation and resupply. Some identifications proved to be NVA units that infiltrated through Cambodia into South Vietnam. In other cases, VC recruits were taken to training areas in Cambodia where new units were formed and later returned to SVN. (These identifications were also derived from headquarters and support elements and units that remained permanently in the enemy base area complexes in Cambodia). Moreover, the study stated that there were four major Communist base areas in the regions of the Chu Pong Mountain (later designated Base Area 732 and subsequently redesignated Base Area 701), the Nam Lyr Mountain (Base Area 740), Lo Go (Base Area 354), and the Bassac-Mekong Corridor (Base Area 704). Three other developing base areas at the time were Tassing in Laos at the northern boundary of Cambodia in the tri-border area (Base Area 609), Snuol and the area to the south and east (Base Area 350), and the Parrot's Beak of Svay Rieng Province (Base Area 367). (See Figure 200-5.)

6. It was noted in the latter part of 1965 that four NVA regiments--the 95B, 88th, 32d and 33d--apparently had planned to launch an offensive in the Western Highlands in November. However, these plans

UNCLASSIFIED

- 28 -

were thwarted by Operation PAUL REVERE IV, a search and destroy operation conducted from 18 October to 31 December in SW Kontum and NW Pleiku Provinces. The four NVA regiments withdrew into sanctuaries in northeastern Cambodia to avoid this operation.

7. The Vietnamese Communists' use of areas of Cambodia for sanctuary and base regions had a profound effect upon Allied operations in 1967. During many of the key battles and campaigns of 1967 (e.g., Operations SAM HOUSTON in Central Highlands and JUNCTION CITY in Northern III CTZ), the enemy relied heavily upon Cambodian territory for strategic and tactical purposes.

8. By October 1967, CICV had listed ten base areas either entirely located within Cambodia or extending into Cambodia from South Vietnam. These base areas were described in detail in CICV studies of April, July and October 1967 which revealed significant Communist activities in the area and the permanent nature of the occupation of Cambodian soil. Furthermore, Vietnamese Communist documents captured in 1967 represented an important source of information concerning the Communists' use of Cambodia. They related that:

a. "Regulations Concerning the Border," dated December 1966 and disseminated in February 1967, provided instructions regarding the construction of bases and their evacuation.

- 29 -

9. 6. The size of ordnance deliveries through Cambodia to the Vietnamese Communists from December 1966 to April 1969 is startling when it is compared with the total enemy external ordnance requirements for the southern half of South Vietnam. For example, in its November 1968 study, MACV J2 estimated that during 1968--the enemy's highest requirement year--that the average monthly ordnance requirements for II, III and IV CTZs were 217.5 short tons. This estimate--which was the result of extensive research--considered expenditures, captured and destroyed items, fair wear and tear and packaging. Measuring in round numbers then and using this estimated monthly requirement against the total ordnance input through Sihanoukville of almost 21,000 tons destined for the Vietnamese Communists, sufficient munitions would have passed through the port to sustain Communist forces' combat activities from December 1966 at the highest levels ever achieved for a total of about eight years. These general estimative computations are based on the premise that all ordnance destined for the enemy would be available to and used by the enemy. However, enemy ordnance losses as a result of Allied cross-border operations and the residual enemy ordnance

- 30 -

C. CAMBODIAN INVOLVEMENT WITH THE VIETNAMESE COMMUNISTS.

1. In 1964 and 1965, scattered reports suggested that the Viet Cong were receiving some local support from Cambodians in the border regions. However, until the end of 1965, there was no indication of official RKG support for the VC/NVA war effort. Prince Sihanouk did provide some moral support for the enemy in 1965 by condemning the role of the US in South Vietnam and by providing medicines for the "victims of American bombing.". On another occasion in 1965, Sihanouk handed over to representatives of the Viet Cong 45 cases of assorted medicines. In late 1965, Cambodia and North Vietnam reportedly signed an agreement to sell 30,000 tons of rice annually to North Vietnam for delivery to VC/NVA forces. Cambodia thereby placed itself in a position that was diametrically opposed to US interests in the area.

2. Although he had reportedly signed an official sales pact with the enemy, Sihanouk was opposed to unofficial trading with the VC/NVA.

Shows the NVA was doing

3. (4) Sihanouk's moral support for the VC/NVA continued in 1966, during which year he raised North Vietnam's commercial representation to government level. In return, North Vietnam guaranteed to respect Cambodia's territorial integrity. In his broadcasts over the Phnom Penh Domestic Radio Service in 1966, Sihanouk admitted that Cambodia was giving physical support to the Vietnamese Communists. In a March broadcast, he stated that wounded VC combatants had been evacuated to Cambodia for rest and would be treated in Cambodian hospitals and then returned home. In another broadcast, the Prince revealed that Vietnamese Communists had requested aid in the form of Cambodian rice and that Cambodia would honor their request. In April, Sihanouk presented 13 tons of dried fish to a member of the Central Committee of the National Liberation Front as a token of friendship. At the presentation ceremony, the Prince spoke of the "unalterable solidarity" of the Cambodian people with the VC "in their struggle for the liberation of the Vietnamese people." Despite these public displays on behalf of the Communists, Sihanouk insisted in 1966 that the VC/NVA did not have permanent installations on Cambodian soil. He also argued that his public statements in favor of the (SLF) and North Vietnam were not inconsistent with Cambodia's official position of neutrality, and he called on the ICC to prove that his country was not harboring concentrations of the enemy.

4. By September 1966, MACV J2 was convinced that Sihanouk

was providing significant national-level support to the enemy. MACV
J2 stated that "evidence shows that Cambodia is selling and delivering
large tonnages of rice to the enemy force." It also noted that the
Cambodian armed forces and their vehicles were providing the primary means
for movement of these tonnages to the enemy. MACV J2 pointed out that
reported national-level contracts from late 1965 through April 1966
between the RKG, Communist China, North Vietnam and the Viet Cong amounted
to 55,000 tons annually. In light of this evidence, it was concluded in
October 1966 that the Cambodian government's position could no longer
be termed neutral, but rather it had become a nonbelligerent ally of the
Vietnamese Communists.

5. MACV J2 asserted in September 1966 that the favorable
attitude shown by Sihanouk toward the Vietnamese Communists created an
atmosphere which permitted nationwide sanction for all peoples of Cambodia
to support the enemy. With this tacit approval, the lower levels of
Cambodia's military, civil service, and local populace became increasingly
involved in supporting the Vietnamese Communists. For example, in addition to the official sales of rice by the RKG, MACV J2 told of reports
indicating that in 1966 Chicom and North Vietnamese officials were purchasing 100,000 to 120,000 tons of rice from Cambodian farmers. According
to MACV J2, numerous reports documented the low-level support. Cambodian
border outposts seldom took any action to restrict movement within or to
eject enemy forces from Cambodia. Cambodian soldiers reportedly shared
facilities and bivouacked with VC/NVA forces. Evidence showed that local
civil authorities gave permission to the enemy to take refuge all along
Cambodia's border.

6. (?) Cambodia's relations with the Vietnamese Communists were greatly enhanced in 1967 when the National Liberation Front and North Vietnam in June formally recognized the "territorial integrity of Cambodia within its present borders." Sihanouk reciprocated by stating that the NLF would "naturally and consequently" have diplomatic representation in Phnom Penh with its representative ranking among ambassadors. Sihanouk also proposed upgrading Hanoi's representation one step to full embassy status. The Vietnamese Communists made as much propaganda value as possible from Sihanouk's recognition. Exchanges of numerous letters and communications between VC and NVN leaders and Sihanouk were highlighted by both Phnom Penh and Hanoi.

7. (?) The formal accreditation ceremony for the NLF took place in Phnom Penh on 30 July 1967. On this occasion, Sihanouk reiterated the "stand of the RKG and the people of Cambodia always to support the South Vietnamese people's just resistance against the US aggressors and for national salvation." North Vietnam's representation was raised to full embassy status at a ceremony in the Royal Palace in Phnom Penh on 27 August, at which time Sihanouk again gave full support to the four-point stand of DRV and the five-point position of the NLF.

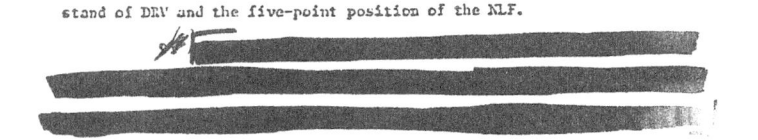

- 34 -

8. Throughout 1968, intelligence reports of varying reliability alluded to an emerging involvement of the Cambodians with VC/NVA forces. From 25 selected reports (17 from fairly and usually reliable sources and 8 from unevaluated sources), 17 reports indicated national level involvement, one provincial level and the remainder significant low level involvement. For example, in June, two members of the tribal autonomy movement (FULRO) cited seven members of FARK who were members on a covert Cambodian committee on contraband with the Viet Cong. Its membership included Major Oum Savouth, Lieutenant Colonel Les Kossem and Captain Lon Non, General Lon Nol's brother. Reports on smuggling activities with the Viet Cong were purportedly submitted to Lon Nol who delivered them to Prince Sihanouk. In October, the same sources described regular Sunday meetings that took place in Les Kossem's home in Phnom Penh between representatives of the NLF, FARK and the RKG. The purpose of the meetings was to discuss the movement of supplies through Cambodia to VC units and to assure that RKG complicity was not exposed. Another source in December attributed overall supervision of arms and rice shipments to Vietnamese Communists troops in the Cambodian border regions to the Hak Ly Trucking Company. Moreover, a MACV J2 compilation of 236 reports covering the period from March 1967 through October 1968 reflected the flow of ordnance from Cambodia into the Republic of Vietnam. The use of Cambodian military trucks in making deliveries to the Viet Cong along the border was described in 29 of these reports.

9. In December 1968, in a message to the Chairman, JCS, COM-USMACV concluded: "Recent evidence has established FARK complicity in the ever-increasing arms traffic to the enemy. The involvement of

UNCLASSIFIED

senior Cambodian officials indicates at least an awareness of this arms movement by the Sihanouk regime."

10. (U) A shift in Cambodian relations with the Vietnamese Communists was noted in 1969. CINCPAC noted that in the first part of the year there were growing indications that the Cambodians were becoming concerned over the continued presence of VC/NVA forces on Cambodian soil and the threat to Cambodian independence which these forces represented. This was heightened by several instances of armed confrontation between FARK and the VC/NVA. According to a 7 March article in Realities Cambodgiennes, Sihanouk admitted that the Communists had infiltrated into Cambodia and that such infiltrations were of great concern to him. At a March press conference, Sihanouk pointed out on a map "the border regions where the Vietnamese are most numerous, in some cases heavily occupying certain zones. /See Annex D2 s.7 Roughly, they stay mainly from east of Mimot to Mondolkiri and Ratanakiri in half-deserted places which the Khmer Army cannot occupy permanently for lack of manpower. They are found east of Sen Monorom, in the valley of Nam Lyr near Bo Kheo and Ba Kham." At a later press conference in March, he stated: "The Communists are playing a dangerous game for themselves and for us."

11. (U) At a 30 April meeting with the press, the Prince described the encroachment of Viet Cong and "Viet Minh" elements in northeast Cambodia as typical tactics of the Asian Communists. In

UNCLASSIFIED

addition, he raised the possibility of severing relations with Hanoi. In a speech on 11 June, Sihanouk related that VC forces were present everywhere along the border. (See Figure 50?-34) He added on 19 June that he had had the NFLSVN sign a paper promising to withdraw from Cambodia as soon as possible, but added that he did not fully trust the written pledge.

12. (X) The Cambodian government began to apply pressure against the VC/NVA, and by April some local Cambodian officials in the border areas had gone so far as to contact their ARVN counterparts to discuss ways in which they could cooperate against the VC. For example, MACV recorded seven meetings which occurred during the week of 7-13 April 1969 between RVN district and provincial officials and Cambodian authorities at four points along the frontier. MACV estimated that FARK efforts against the VC/NVA were primarily designed to re-establish Cambodian sovereignty in areas which the VC/NVA were willing to vacate, to impede further infiltration, and to restrict VC/NVA activities to specified enclaves. The change in the Cambodian attitude, according to MACV, was partly the result of heavier fighting along the border and the extent to which the Communist troops had made increasing use of their bases in the southern half of Cambodia. The Cambodians also believed that the Communists were supporting the local rebels.

13. (U) For their part, the Communists tried to maintain cordial relations. The Cambodian actions did not seriously threaten Communist bases in Cambodia, for the small Cambodian army posed no real threat. However, Sihanouk did have one trump card -- his ability to embargo supplies and munitions passing to the VC/NVA through Sihanoukville. In May, he employed this embargo because of the deteriorating relations and

UNCLASSIFIED

- 37 -

cut off the flow of munitions and other supplies moved via the depots at Kompong Speu and Lovek. The stoppage applied to the whole system. Negotiations between the RKG and the Provisional Revolutionary Government commenced when PRG President Huynh Tan Phat visited Phnom Penh from 30 June to 2 July, but the embargo continued and lasted until mid-September. At that time, Sihanouk traveled to Hanoi to attend the funeral of Ho Chi Minh. While he was there, a meeting was arranged, apparently by Chou En-lai, between Sihanouk and VC and NVN leaders. According to a MACV assessment, the meeting resulted in a quid pro quo arrangement whereby Sihanouk would release the arms and the VC/NVA would confine their activities to specified enclaves, stop aiding the Khmer Rouge and Khmer Loeu insurgents, and stop harassing the local population.

14. ~~S, C~~ Thus, in the latter part of 1969, relations again began to improve. A marked decrease in the number of incidents between the Cambodian army and the VC/NVA during the last five months of the year was noted. Moreover, a trade agreement between the RKG and the PRG was signed on 25 September for the trade of non-military goods. This agreement and Sihanouk's move to dissolve the International Control Commission in late 1969 were seen as attempts by Sihanouk to maintain good relations with Communist China, North Vietnam and the PRG.

15. ~~S, C~~ With the departure of Sihanouk for France in early January 1970, a new phase in Cambodian/Communist relations opened under acting Prime Minister Sirik Matak, who tried to pressure the Vietnamese Communists into reducing their military presence in Cambodia. A mid-January meeting between Matak and Tran Buu Kiem, a high-ranking PRG official, demonstrated the ill will between the Lon Nol-Sirik Matak government and the Vietnamese

UNCLASSIFIED

Communists. Matak reportedly accused the Communists of killing many of
his countrymen at a time when Cambodia was aiding the Viet Cong and
threatened to take military action against them. Kiem allegedly responded with a warning that any such action would result in hard fighting.
Matak then reportedly ordered FARK to destroy Communist installations,
including medical facilities in Svay Rieng Province. The Cambodian
Army also continued its campaign in the northeast to try to regain some
control over Communist and insurgent-held territory.

Prime Minister Lon Nol was concerned with Viet Cong support
of local insurgents. An article attributed to him which appeared in an
official journal in March gave an unusually detailed account of Cambodian
guerrilla groups based in Viet Cong-controlled areas of South Vietnam
that had been crossing into three southeastern border provinces to spread
anti-government propaganda. The article alleged that these groups,
totaling about 450 men, were led by long-time members of the Cambodian
Communist Party, and apparently planned to extend their operations westward.

16. (a) The Cambodian government placed a 15 March deadline for
VC/NVA forces to leave Cambodia. Pressure continued to build when on
11 March the North Vietnamese and PRG Embassies were sacked by thousands
of students. Anti-Viet Cong activities also took place in several
provinces. Prior to the 15 March deadline, a compromise was reported
between the RKG and North Vietnam, and the parties agreed to meet on

- 39 -

16 March to discuss the settlement of problems.

17. ~~(...)~~

~~[redacted]~~

On 9 April, [redacted] reported that Cambodian officials had taken measures to close the Tonle Kong River to Vietnamese Communist traffic. The waterway had been used since 1966 to transport supplies to Laos.

18. In response to these moves by the RKG, the Vietnamese Communists began to move militarily in Cambodia. On 17 April, they attacked government outposts all along the frontier. The most sustained action occurred in Svay Rieng Province, while Communist military activity rose substantially elsewhere along the border. Phnom Penh then substantially reinforced the military districts east of the capital. Later in April, the Communists continued their intensified military pressure, the heaviest action being concentrated in Kandal and Takeo Provinces south of Phnom Penh. The situation was unstable and was rapidly deteriorating in the countryside. The Communists continued their attacks against government positions and key lines of communication in widely separated areas of eastern Cambodia. Phnom Penh was taking an increasingly tough line with the Communists.

19. 2. The Communists at this time also began to devote

- 40 -

considerable attention to forming a "Cambodian liberation movement"
and to expand "liberated areas" in Cambodia. With the fall of Stung
Treng on 15 May, the Communists acquired control over the Se Kong-
Mekong waterway from the Laos border to the outskirts of Kompong Cham.
Lon Nol's government position deteriorated in the countryside in the
face of increasing military operations. According to a late August
MACV assessment, a study of enemy activities in Cambodia suggested that
Hanoi probably established five near-term objectives in Cambodia: (a)
to re-establish a secure LOC system in Cambodia, (b) to achieve freedom
of movement in eastern Cambodia, (c) to organize a guerrilla base to
develop a Khmer insurgency movement, (d) to isolate Phnom Penh and other
population centers, and (e) to consolidate control over the Tonle Sap
area. To achieve these objectives, the enemy deployed an estimated
53,000 men in or near Cambodia.

20. M. 22. The Communists continued to harass the provincial
capitals and Cambodian LOC's from September to the end of the year.
MACV estimated from observing the pattern of Communist activity that the
enemy probably was using a protracted war strategy. This stragegy was
probably designed to (a) cripple and demoralize FARK forces, (b) provide
a screen for logistical activities, (c) create insecurity in the countryside, (d) isolate the populace from the government, (e) develop an
infrastructure, and (f) wreck the national economy.

UNCLASSIFIED

D.1. LOGISTICS.

1.(u) From the early 1960's into 1970, the MACV assessments concerning the Vietnamese Communist logistical system which supported Communist forces in South Vietnam were consistent; they indicated unerringly that enemy forces in northern South Vietnam received their supplies across the DMZ and through Laos while those in southern South Vietnam received the bulk of their supplies from Cambodia once MARKET TIME Operations constituted an effective barrier to direct sea deliveries by the North Vietnamese by mid-1966.

2. Significantly, in August 1964, an official Vietnamese service intelligence study discussed Viet Cong supply routes. believed that the evidence in the study was convincing enough to regard the report as probably true. The study in part stated: "The GVN realized the importance of the supply routes from Cambodia in relation to the increased fighting capacity of the VC in South Vietnam. At present because of the heightened development of their forces, VC demands for equipment have increased many times over, and one of their principal concerns, therefore, is logistical operations from outside Vietnam. These operations can be implemented in three ways: by sea routes from North Vietnam, by land routes from Laos, and by land and sea routes from Cambodia. Because of the increased patrol operations by the GVN Navy and because VC transports can be uncovered very easily, the VC have used the sea routes only with the greatest of care. Land routes through Laos, with their many mountains and other obstacles, have not been very conducive to large-scale supply movements, and have, as a result, been used principally for infiltration of cadres to South Vietnam.

Therefore, because of the inadvisability of using these routes from
North Vietnam and Laos very frequently, the VC have been compelled to
rely almost exclusively on the routes through Cambodia for their
supplies." Using the ~~approach mentioned above, this section will~~
consider the logistical situation of Communist forces in southern South
Vietnam and the concomitant role of the sea route from North Vietnam,
the overland route from Laos and the sea routes and the land routes
from Cambodia or more simply the Cambodian routes. (See Figure 404-S.)

3. The Sea Route from North Vietnam.

a. Sea infiltration of materials and selected cadres
into RVN by steel-hulled trawlers capable of carrying more than 100 tons
of cargo probably started in the late 1950's. By 1960, the North
Vietnamese had established an official infiltration unit, designated
Group 125, which in time came under the direct control of the Military
High Command in Hanoi. There were numerous reports from detainees and
ralliers of extensive infiltration shipments being made to RVN in 1963,
1964 and early 1965. However, after the Vung Ro Bay (Phu Yen Province,
RVN) incident of 16 February 1965, which involved the destruction of a
Group 125 trawler loaded with arms and ammunition, a sharp decline in
trawler shipments occurred as a counter-sea-infiltration effort, MARKET
TIME, began operating in March 1965 to seal the RVN coast.

b. With the development and refinement of Operation
MARKET TIME in 1965 and 1966, the enemy's ability to be supplied by sea
infiltration of materiel decreased drastically. COMUSMACV estimated
that in 1965 approximately 70 percent of the NVA/VC resupply was accomplished
by maritime infiltration. In mid-1966, COMUSMACV considered that the

UNCLASSIFIED

success of MARKET TIME operations was such that not more than ten percent of NVA/VC logistic support came by sea. By the end of 1966, COMUSMACV was able to state that there was no credible evidence of any significant infiltration of enemy troops or supplies by sea since November 1965. All indications were that Operation MARKET TIME had interdicted this avenue of resupply and continued effectively to do so up through 1970. ~~Moreover, the success of MARKET TIME was further reflected by the absence of reports since the end of 1966 indicating traffic in arms reaching the Republic of Vietnam directly from the South China Sea and by the fact that enemy real service groups after 1965 became satellited on the Cambodian border regions.~~

c. Overall, since ~~February 1965, 28 trawlers have~~ been ~~detected attempting to infiltrate~~ supplies into South Vietnam. (See ~~Figure 800-S.)~~

FIGURE 800-S

TRAWLER INFILTRATION INCIDENTS
(February 1965 – December 1970)

YEAR	ATTEMPTS	FAILURES ABORTED	FAILURES DESTROYED/CAPTURED	SUCCESSES
1965	2		1	–
1966	3	1	2	–
1967	3	–	3	–
1968	5	2	3	–
1969	3	3		–
1970	12	10	1	1
TOTAL:	28	17	10	1

SOURCE: Rpt (S), ACS1, DI-ISD, 4 November 1970 (updated Dec 1970), Subject: Trawler Infiltration Incidents.

UNCLASSIFIED

VC/NVA BASE AREAS IN CAMBODIA

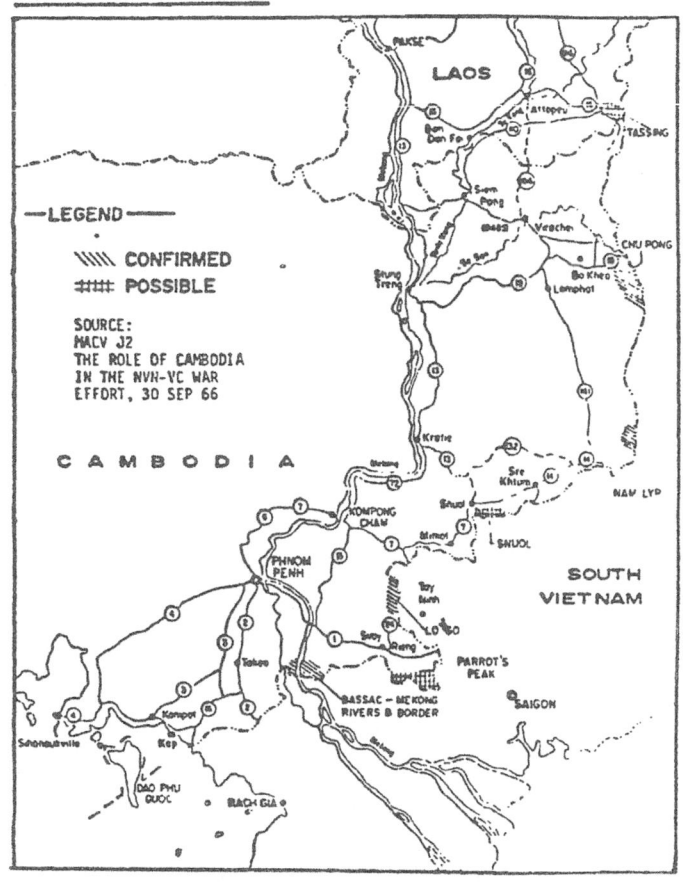

LEGEND

\\\\\\ CONFIRMED
POSSIBLE

SOURCE:
MACV J2
THE ROLE OF CAMBODIA
IN THE NVN-VC WAR
EFFORT, 30 SEP 66

UNCLASSIFIED

Summary: 1953-1970

A. POLITICAL.

1. From November 1953, when Cambodia became independent, until March 1970, when Prince Sihanouk was overthrown, the primary concern of Sihanouk's foreign policy was to preserve the independence and territorial integrity of Cambodia. In his view, the chief threats to those objectives were posed by Thailand and Vietnam. For several centuries, these countries had encroached on Khmer Territory, and Prince Sihanouk feared that they still had designs on Cambodia.
2. At the time of gaining independence, Cambodia was small, weak, and in a vulnerable position. Consequently, to fulfill his foreign policy objective of preserving Cambodia's independence and territorial integrity, Sihanouk was faced with the decision of joining one of the two great power blocs or of assuming a neutral position. By late 1954, he decided on the latter course, proclaiming in December that Cambodia would remain nonaligned and

would adhered to an official policy of neutrality.
3. For the next decade, Sihanouk held to this neutralist course. In September of 1957, he had Cambodia's National Assembly pass the Neutrality Act, which declared that Cambodia would abstain from aggression, while reserving the right to self-defense. Three years later, Sihanouk proposed to the United States that a neutral zone composed of Laos and Cambodia be established and guaranteed by the major power blocs. In both 1962 and 1963, be proposed a multinational moratorium conference to guarantee Cambodia's independence, neutrality, territorial integrity and sovereignty.
4. In late 1963, although still working within the framework of an official foreign policy of neutrality, Sihanouk began his drift to the left, as evidenced by his termination of US aid in November. The reasons for this shift rested in Sihanouk's reappraisal of the US role in Southeast Asia, his belief of an emerging hegemony of Communist China, and his fears of an intensified Vietnam War, which be believed North Vietnam

would eventually win. His primary concern remained the survival of Cambodia, and he believed that a movement toward the Communist Bloc would be the best guarantee of Cambodia's survival. Cambodia's swing to the left continued until Sihanouk's ouster, although there were periodic retrogressions during times when Sihanouk desired to keep some semblance of neutrality for the purpose of appeasing growing Western apprehensions.

5. During 1965, Sihanouk turned abruptly from the United States, and on 3 May he announced that he was severing diplomatic relations with the US. It was noted in October 1965 that Cambodia, having departed from its enunciated policy of neutralism, had become aligned in a sense to the Communist Bloc. Sihanouk was making efforts to ingratiate himself and his nation with the leaders of "the wave of the future" in order to preserve his nation. One of the steps taken by Sihanouk was to accept a Chicom military aid program and to enter into a Chinese-Cambodian cultural agreement. In his public address, in 1965, Sihanouk emphasized

Cambodia's friendship with China, which, he said, had promised to stand behind his country in its struggles in Southeast Asia.
6. This trend continued in 1966, and MACV J2 noted a growing Chicom influence in Cambodia. Military intelligence in Saigon felt that Sihanouk's motives had not changed and that he was orienting his country toward Red China as a result of his belief that it was the best way to safeguard Cambodia's sovereignty and territorial integrity with himself in power. Sihanouk remained chary in his relations with South Vietnam and Thailand, using the occasion of periodic border incidents to warn of the expansionistic tendencies of these neighbors. Despite his close association with the Communist Bloc, Sihanouk argued that his country was still neutral. He offered on several occasions the proposal that the International Control Commission (ICC) be expanded so that it could effectively patrol the border. The Soviets and Chinese objected to the proposal, however.
7. The first major retrenchment in Cambodia's close association with China

occurred in 1967. Foreign military observers in Phnom Penh noted that Sihanouk was becoming concerned both with China's growing influence in Cambodia and with the militant activities of the domestic Cambodian Communist rebels, the Khmer Rouge. In order to keep a balance between the Western and Eastern influences in his country, especially after the Khmer Rouge-inspired rebellion in Battambang Province in early 1967. At one point, he threatened to close his embassy in Peking. However, after the receipt of a conciliatory letter from Premier Chou En-Lai in September, Sihanouk softened his hard stand against Peking. During the year, however, he persisted in his chilly attitude toward the US and the Allies. He adamantly denied US claims that the enemy was using Cambodian territory, arguing that these claims indicated the Allies' intentions to expand the war into Cambodia. However, to give evidences of his neutral posture, Sihanouk again extended offers to expand the role of the ICC to guarantee the inviolability of Cambodian soil. Also, he allowed American newsman to

tour Cambodia without much restriction. However, following the discovery and report by newsmen in November 1967 of a Vietnamese Communist base camp in the Mimot area, Sihanouk quickly denied the story and reused future examination of the border area by Western journalists.
8. There was a temporary swing away from further rapprochement with the Vietnamese Communists in 1968, for Sihanouk was becoming aware of the extent of their use of Cambodian territory. The role of Cambodia as a supply base for the enemy in the Tet Offensive tended to exacerbate the Prince's quest for guaranteed neutrality for Cambodia. Sihanouk was said to be particularly upset and enraged when he saw the US documentation of Viet Cong activities in Cambodia. This raised hopes of a closer alignment for Cambodia with the United States. Nevertheless, through the course of the year, Sihanouk kept denouncing Americans for not wanting peace in Vietnam, and he leveled ascorbic attacks against the US as a result of border incidents. Relations with Communist China remained unchanged in 1965, although Sihanouk

did express concern over possible Chinese denunciation of Cambodia in the future. The Prince expressed hopes that upon leaving SVN, US would remain in Thailand and the Philippines to offset Chinese influence.

9. During 1969, Prince Sihanouk's foreign policy tactics included vacillating between Vietnamese Communists and the United States. Three primary factors influenced his political maneuvering:

   A) Sihanouk realized that the VC/NVA were not winning the war and that the struggle had become much longer than originally anticipated;

   B) Allied military success in South Vietnam had caused the development of large, permanently garrisoned VC/NVA enclaves in Cambodia, over which Cambodia had limited or no sovereignty;and

   C) Sihanouk feared the indigenous Communist insurgency in Cambodia and the aid which the VC, NVN and Communist China were giving it.

10. As a result of these factors, Cambodia drew closer to the US in 1969. In March, Prince Sihanouk admitted that the VC/

NVA were infiltrating through Cambodia and occupying parts of his country. Later he spoke for the first time of possibly severing relations with Hanoi. As relations with the US improved in the spring, Sihanouk moved to cut off the flow of munitions and other supplies to the VC/NVA by placing an embargo on the transshipment of arms and supplies from the depots at Kampong Speu and Lovek. This embargo lasted until mid-September 1969, when Sihanouk apparently reached an understanding with the leaders of the Viet Cong and North Vietnamese during a meeting in Hanoi.

11. The principal means of Allied pressure for a change in Cambodian foreign policy was the threat of hot-pursuit or cross-border operations directed against VC/NVA installations in the border regions. Moreover, the possibility of improved US relations carried with it the prospect of significant aid assistance, which would help to reduce Cambodia's economic strains.

12. The severe economic troubles caused Sihanouk to install Lon Nol as Prime Minister on 12 August 1969, charging him with the responsibility of rectifying

the economic problems. Since Lon Nol was almost continually absent because of personal reasons and state visits, Deputy Prime Minister Prince Sisowath Sirik Matak, for all practical purposes, served as Prime Minister. Matak's relations with Sihanouk deteriorated during the period over the question of how to implement necessary economic reforms. Moreover, it was reported that between August and the end of 1969, Sihanouk made three attempts to dissolve the government through political maneuvering. However, he met opposition- the first time in a decade that he had faced concerted resistance to his policies.

13. With both Sihanouk and Lon Nol out of Cambodia in early 1970, Matak was in charge of the government for about a month and a half. When Lon Nol returned to Cambodia on 18 February 1970, the Royal Khmer Government (RKG) expressed new concern about several fundamental problems with the Vietnamese Communists. On 11 March 1970, the North Vietnamese and the Provisional Revolutionary Government (PRG) embassies were sacked by thousands of students. These

assaults were preceded by anit-VC/NVA demonstrations in several provinces. These attacks received the unanimous support of the Cambodian legislature, which passed a declaration requesting the government to take all measures necessary to solve the problem of enemy infiltration. Finally, on 18 March 1970, the National Assembly unanimously voted Sihanouk out of office as Head of State. Subsequently, Sihanouk was stripped of his military and political party titles, and the army was ordered "to crush by means of arms and all actions which Prince Norodom Sihanouk may provoke."

14. Seeking refuge in Peking, Sihanouk announced in a series of radio broadcasts from China his intention to lead the struggle to oust the Lon Nol-Sirik Matak government. He announced the formation of an exile government- the Royal Government of National Union. He also told of plans to organize the newly-formed National United Front of Kampuchea at the grassroots level throughout Cambodia. The North Vietnamese and Viet Cong, faced with a threat to their Cambodian sanctuaries and resupply haven, moved with great

speed following the 18 March dismissal of Sihanouk. They verbally supported Sihanouk, hardly denounced the new regime, withdrew their embassy personnel, suspended diplomatic relations, and tried to create and exploit civil chaos. The Chinese Communists applied pressure on the Lon Nol government by providing Sihanouk with a forum and by offering aid for the liberation of Cambodia. Meanwhile, the Soviet Union adopted a posture of extreme caution toward events in Cambodia.

15. Saigon lost little time in capitalizing militarily on Sihanouk's ouster. In early April 1970, ARVN troops began the first of a series of forays across the Cambodian border into Communist base areas. Concurrently, the Communists began stepping up their attacks on major towns and lines of communication in areas south and east of Phnom Penh. The subsequent massive Allied cross-border attacks against enemy military sanctuaries in May and June tended to put an official and long-term stamp on the emerging military relationship between Phnom Penh and Saigon, especially in view of the fact that continuous

(Army of the Republic of Vietnam-ARVN) support would be necessary until the Cambodian Army could assume the bulk of the combat burden.

16. With Prince Sihaounk's ouster, the most important cause of discord in Thai-Cambodian relations was removed. The meeting between Foreign Ministers Thanat and Yem Sambaur in Bangkok on 3 May 1970 resulted in agreement on the following points: (a) South Vietnamese troops were operating in Cambodia with the agreement of the Government of Cambodia (GOC); (b) diplomatic relations were to be established; (c) the GOC agreed to protect the lives and goods of Vietnamese residing in Cambodia and to take all necessary measures to protect the security of Vietnamese who desired to stay in Cambodia; and (d) each country pledged itself to respect the border of the other.

17. As the year ended, a 255-million-dollar MAP (Military Aid Program) authorization for Cambodia was approved by the US Congress on 22 December 1970. It contains the provision that no US ground combat troops could be introduced into Cambodia, unless the President deemed

such a step necessary to promote the safe and orderly withdrawal of American troops from South Vietnam or to obtain the release of Americans held as prisoners of war.

B. <u>VC/NVA USE OF CAMBODIA</u>.

1. The Vietnamese Communists' use of Cambodian territory over the year was poignantly portrayed in the following comments:

   (a) On 24 February 1968 at a news conference, General Westmoreland said:
   "The enemy has ignored the neutrality of Cambodia. . . and has taken advantage of this situation and has used the border areas of Cambodia for purposes of infiltration, supply and troop rehabilitation and training. Even in the populated areas of Cambodia adjacent to South Vietnam, a smuggling operation of major

proportions has been conducted to supply the Viet Cong forces with arms, ammunition and medical supplies."

(b) Then, on 30 June 1970, in commenting about the Allied cross-border operations in Cambodia, President Nixon said:

"The prospect suddenly loomed of Cambodia's becoming virtually one large base area for anywhere in South Vietnam along the 600 miles of Cambodian frontier. . . The possibility of a grave threat to our troops in South Vietnam was rapidly becoming an actuality."

2. Even though little information was available in 1964 concerning Vietnamese Communist installations and supply centers on Cambodian soil, MACV estimated that the Viet Cong were probably using Cambodian territory for recuperation, regroupment and resupply. Moreover, the importance of Cambodia to the enemy was underscored in a document taken from a VC prisoner who in December 1963 headed the Psychological Warfare Section, Central Guerrilla Headquarters. It

stated that the neutrality of Cambodia and Laos facilitates the expansion of secure base areas, without which "we cannot develop our mission." Five areas along the Cambodian-Republic of Vietnam border were suspected of being Viet Cong sanctuaries.
3. By the end of 1965, it was reported that the evidence showing the Vietnamese Communists' reliance on Cambodian territory had become indisputably conclusive. This evidence revealed the Communists used their Cambodian sanctuary regions for the following reasons:
   a. As a refuge when pursued by American or ARVN forces.
   b. As a training or hospital area.
   c. As a staging area.
   d. As a logistical area.
   e. As safe infiltration or transmit routes.
   f. As a communications zone.

4. The importance of Cambodia to Communist forces for supporting their offensive actions in South Vietnam became abundantly clear in October 1965, when three North Vietnamese Army (NVA)

Regiments staged outside of Cambodia for the Ia Drang Valley battle with US forces. Several of the more than 100 prisoners captured in this battle told of their passage through and training in Cambodia prior to the engagement and the movement of supplies from Cambodia to the battle area. Moreover, intelligence indicated that small groups of Communist forces had taken refuge in Cambodia following the Ia Drang Campaign.

5. In a study in September 1966, MACV J2 stated that, since the latter part of 1965, there had been a marked increase in reported identifications of NVA and VC units in Cambodia. This trend was attributed to the increased and practically unrestricted use of Cambodia by units that moved across the border for reorganization, rehabilitation and resupply. Some identifications proved to be NVA units that infiltrated through Cambodia into South Vietnam. In other cases, VC recruits were taken to training areas in Cambodia where new units were formed and later return to SVN. These identifications were also derived from

headquarters and support elements and units that remained permanently in the enemy base area complexes in Cambodia. Moreover, the study stated that there were four major Communist base areas in the regions of the Chu Pong Mountain (later designated Base Area 732 and subsequently re-designated Base area 701), the Nam Lyr Mountain (Base Area 740), Lo Go (Base Area 354), and the Bassac-Mckong Corridor (Base Area 704). Three other developing base areas at the time were Tassing in Laos at the northern boundary of Cambodia in the tri-border area (Base Area 609), Snuol and the area to the south and east (Base Area 350), and the Parrot's Beak of Svay Rieng Province (Base Aree 367).

6. It was noted in the latter part of 1965 that four NVA regiments – the 95B, 88th, 32nd, and 33rd– apparently had planned to launch an offensive in the Western Highlands in November. However, these plans were thwarted by Operation PAUL REVERE IV, a search and destroy operation conducted from 18 October to 31 December in SW Kontum and NW Pleiku Provinces. The four NVA regiments withdrew into sanctuaries in

north-eastern Cambodia to avoid this operation.

7. The Vietnamese Communists' use of areas of Cambodia for sanctuary and base regions had a profound effect upon Allied operations in 1967. During many of the key battles and campaigns of 1967 (e.g. Operations SAM HOUSTON in Central Highlands and JUNCTION CITY in Northern III CTZ), the enemy relied heavily upon Cambodian territory for strategic and tactical purposes.

8. By October 1967, CICV had listed ten base areas either entirely located within Cambodia or extending into Cambodia from South Vietnam. These base areas were described in detail in CICV studies of April, July and October 1967, which revealed significant Communists activities in the area and the permanent nature of the occupation of Cambodian soil.

9. The size of ordnance deliveries through Cambodia to the Vietnamese Communists from December 1966 to April 1969 is startling when it is compared with the total enemy external ordnance requirements for the south half of South Vietnam. For example, in its

November 1968 study, MACV J2 estimated that during 1968 - the enemy's highest requirement year - that the average monthly ordnance requirements for II, III and IV CTZs were 217.5 short tons. This estimate, which was the result of extensive research, considered expenditures, captured and destroyed items, fair wear and tear, and packaging. Measuring in round numbers, then, and using this estimated monthly requirement against the total ordnance input through Sihanoukville of almost 21,000 tons destined for the Vietnamese Communists, sufficient munitions would have passed through the port to sustain Communist forces' combat activities from December 1966 at the highest levels ever achieved for a total of about eight years. These general estimative computations are based on the premise that all ordnance destined for the enemy would be available to and used by the enemy.

ARMS AND AMMUNITION SHIPMENTS

SHIP	REGISTRY	ARR DATE	TONNAGES PROBABLE	POSSIBLE
YOU HAO	CHICOM	OCT 66		250
HE PING	CHICOM	DEC 66	450	
HE PING	CHICOM	MAR 67		40
YOU YI	CHICOM	OCT 67		805
FO SHAN	CHICOM	DEC 67	11	
YOU YI	CHICOM	JAN 68	3,848	
WU XI	CHICOM	MAR 68	3,000	
SVOBODA	SOVIET	JUL 68	40	
FO SHAN	CHICOM	AUG 68		2,000
PARTIZANSKAYA SLAVA	SOVIET	OCT 68	55	
BEREZOVKA	SOVIET	DEC 68	76	
LI MING	CHICOM	JAN 69		4,500
HUANG SHI	CHICOM	MAR 69		2,000
YOU YI	CHICOM	JUL 69		
		TOTALS	7,480	9,595

SOURCE: Article (S), PACOM, Intelligence Digest, No. 18-69, 29 August 1969, Subject: The Logistical Importance of Cambodia to the Enemy

UNCLASSIFIED

C. CAMBODIAN INVOLVEMENT WITH THE VIETNAMESE COMMUNISTS.

1. In 1964 and 1965, scattered reports suggested that the Viet Cong were receiving some local support from Cambodians in the border regions. However, until the end of 1965, there was no indication of official RKG support for the VC/NVA war effort. Prince Sihanouk did provide some moral support for the enemy in 1965 by condemning the role of the US in South Vietnam and by providing medicines for the "victims of American bombing." On another occasion in 1965, Sihanouk handed over to representatives of the Viet Cong 45 cases of assorted medicines. In late 1965, Cambodia and North Vietnam reportedly signed an agreement to sell 30,000 tons of rice annually to North Vietnam for delivery to VC/NVA forces. Cambodia thereby placed itself in a position that was diametrically opposed to US interests in the area.
2. Although he had reportedly signed an official sales pact with the enemy, Sihanouk was opposed to unofficial trading with the VC/NVA.

3. Sihanouk's moral support for the VC/NVA continued in 1965, during which year he raised North Vietnam's commercial representation to government level. In return, North Vietnam guaranteed to respect Cambodia's territorial integrity. In his broadcasts over the Phnom Penh Domestic Radio Service in 1966, Sihanouk admitted that Cambodia was giving physical support to the Vietnamese Communists. In a March broadcast, he stated that wounded VC combatants had been evacuated to Cambodia for rest and would be treated in Cambodian hospitals and then returned home. In another broadcast, the Prince revealed that Vietnamese Communists had requested aid in the form of Cambodian rice and that Cambodia would honor their request. In April, Sihanouk presented 13 tons of dried fish to a member of the Central Committee of the National Liberation

Front (NLF) as a token of friendship. At the presentation ceremony, the Prince spoke of the "unalterable solidarity" of the Cambodian people with the VC "in their struggle for the liberation of the Vietnamese people." Despite these public displays on behalf of the Communists, Sihanouk insisted in 1966 that the VC/NVA did not have permanent installations on Cambodian soil. He also argued that his public statements in favor of the National Liberation Front and North Vietnam were not inconsistent with Cambodia's official position of neutrality, and he called on the ICC to prove that his country was not harboring concentrations of the enemy.

4. By September 1966, MACV J2 was convinced that Sihanouk was providing significant national-level support to the enemy. MACV J2 stated that "evidence shows that Cambodia is selling and delivering large tonnages of rice to the enemy force." It also noted that the Cambodian armed forces and their vehicles were providing the primary means for movement of these tonnages to the enemy. MACV J2 pointed out that reported national-level contracts from

late 1965 through April 1966 between the RKG, Communist China, North Vietnam and the Viet Cong amounted to 55,000 tons annually. In light of this evidence, it was concluded in October 1966 that the Cambodian government's position could no longer be termed neutral, but rather it had become a nonbelligerent ally of the Vietnamese Communists.

5. MACV J2 asserted, in September 1966, that the favorable attitude shown by Sihanouk toward the Vietnamese Communists created an atmosphere that permitted nationwide sanction for all peoples of Cambodia to support the enemy. With this tacit approval, the lower levels of Cambodia's military, civil service, and local populace became increasingly involved in supporting the Vietnamese Communists. For example, in addition to the official sales of rice by the RKG, MACV J2 told of reports indicating that, in 1966, Chicom and North Vietnamese officials were purchasing 100,000 to 120,000 tons of rice from Cambodian farmers. According to MACV J2, numerous reports documented the low-level support. Cambodian border outposts seldom took

any action to restrict movement within or to eject enemy forces from Cambodia. Cambodian soldiers reportedly shared facilities and bivouacked with VC/NVA forces. Evidence showed that local civil authorities gave permission to the enemy to take refuge all along Cambodia's border.

6. Cambodia's relations with the Vietnamese Communists were greatly enhanced in June 1967 when the National Liberation Front (NLF) and North Vietnam formally recognized the "territorial integrity of Cambodia within its present borders." Sihanouk reciprocated by stating that the NLF would "naturally and consequently" have diplomatic representation in Phnom Penh with its representative ranking among ambassadors. Sihanouk also proposed upgrading Hanoi's representation one step to full embassy status. The Vietnamese Communists made as much propaganda value as possible from Sihanouk's recognition. Exchanges of numerous letters and communications between VC and NVA leaders and Sihanouk were highlighted by both Phnom Penh and Hanoi.

7. The formal accreditation ceremony for the NLF took place in Phnom Penh on 30 July 1967. On this occasion, Sihanouk reiterated the "stand of the RKG and the people of Cambodia always to support the South Vietnamese people's just resistance against the US aggressors" and for "national salvation." North Vietnam's representation was raised to full embassy status at a ceremony in the Royal Palace in Phnom Penh on 27 August, at which time Sihanouk again gave full support to the four-point stand of DRV and the five-point position of the NLF.

8. Throughout 1968, intelligence reports of varying reliability alluded to an emerging involvement of the Cambodians with VC/NVA forces. From 25 selected reports (17 from fairly and usually reliable sources and 8 from unevaluated sources), 17 reports indicated national-level involvement, one provincial-level and the remainder significant low-level involvement. For example, in June, two members of the tribal autonomy movement (FULRO) cited seven

members of the Cambodian Army (FARK) who were members of a covert Cambodian committee on contraband with the Viet Cong. Its membership include Major Oum Savouth, Lieutenant Colonel Les Kossem and Captain Lon Non, General Lon Nol's brother. Reports on smuggling activities with the Viet Cong were purportedly submitted to Lon Nol who delivered them to Prince Sihanouk. In October 1968, the same sources described regular Sunday meetings that took place in Les Kossem's home in Phnom Penh between representatives of the NLF, FARK and the RKG. The purpose of these meetings was to discuss the movement of supplies through Cambodia to VC units and to assure that RKG complicity was not exposed. Another source in December attributed overall supervision of arms and rice shipments to Vietnamese Communists troops in the Cambodian border regions to the Hak Ly Trucking Company. Moreover, a MACV J2 compilation of 236 reports covering the period from March 1967 through October 1968 reflected the flow of ordnance from Cambodia into the Republic of Vietnam. The use of Cambodian military trucks

in making deliveries to the Viet Cong along the border was described in 29 of these reports.
9. In December 1968, in a message to the Chairman, JCS, COMUSMACV concluded: "Recent evidence has established FARK complicity in the ever-increasing arms traffic to the enemy. The involvement of senior Cambodian officials indicates at least an awareness of this arms movement by the Sihanouk regime."
10. A shift in Cambodian relations with the Vietnamese Communists was noted in 1969. CINPAC noted that in the first part of the year there were growing indications that the Cambodians were becoming concerned over the continued presence of VC/NCA forces on Cambodian soil and the threat to Cambodian independence that these forces represented. This was heightened by several instances of armed confrontation between FARK and the VC/NVA. According to a 7 March article in <u>Realties Cambodgiennes</u>, Sihanouk admitted that the Communists had infiltrated into Cambodia and that such infiltrations were of great concern to him. At a March press conference, Sihanouk pointed out on a map "the

border regions where the Vietnamese are most numerous, in some cases heavily occupying certain zones. Roughly, they stay mainly from east of Mimot to Mondolkiri and Ratanakiri for lack of manpower. Thet are found east of Sen Monorom, in the valley of Nam Lyr near Bo Kheo and Ba Kham." At a later press conference in March, he stated, "The Communists are playing a dangerous game for themselves and for us."

11. At a 30 April meeting with the press, the Prince described the encroachment of Viet Cong and "Viet Minh" elements in northeast Cambodia as typical tactics of the Asian Communists. In addition, he raised the possibility of severing relations with Hanoi. In a speech on 11 June, Sihanouk related that VC forces were present everywhere along the border. He added on 19 June that he had had the NFLSVN sign a paper promising to withdraw from Cambodia as soon as possible, but added that he did not fully trust the written pledge.

12. The Cambodian government began to apply pressure against the VC/NVA, and, by April, some local Cambodian officials in the border areas had

gone so far as to contact their ARVN counterparts to discuss ways in which they could cooperate against the VC. For example, MACV recorded seven meetings that occurred during the week of 7-13 April 1969 between RVN district and provincial officials and Cambodian authorities at four points along the frontier. MACV estimated that FARK efforts against the VC/NVA were primarily designed to re-establish Cambodian sovereignty in areas which the VC/NVA were willing to vacate, to impeded further infiltration, and to restrict VC/NVA activities to specified enclaves. The change in Cambodian attitude, according to MACV, was partly the result of heavier fighting along the border and the extent to which the Communist troops had made increasing use of their bases in the southern half of Cambodia. The Cambodians also believed that the Communists were supporting the local rebels.

13. For their part, the Communists tried to maintain cordial relations. The Cambodian actions did not seriously threaten Communist bases in Cambodia, for the small Cambodian army posed no

real threat. However, Sihanouk did have one trump card - his ability to embargo supplies moving and munitions passing to the VC/NVA through Sihanoukville. In May, he employed this embargo because of the deteriorating relations and cut off the flow of munitions and other supplies move via the depots at Kompong Speu and Lovek. The stoppage applied to the whole system. Negotiations between the RKG and the Provisional Revolutionary Government commenced when PRG President Huynh Tan Phat visited Phnom Penh from 30 June to 2 July, but the embargo continued and lasted until mid-September. At that time, Sihanouk traveled to Hanoi to attend the funeral of Ho Chi Minh. While he was there, a meeting was arranged, apparently by Chou En-Lai, between Sihanouk and VC and NVA Leaders. According to a MACV assessment, the meeting resulted in a quid pro quo arrangement whereby Sihanouk would release the arms and the VC/NVA would confine their activities to specified enclaves, stop aiding the Khmer Rouge and Khmer Loeu insurgents, and stop harassing the local population.

14. Thus, in the latter part of 1969, relations again began to improve. A marked decrease in the number of incidents between the Cambodian army and the VC/NVA during the last five months of the year was noted. Moreover, a trade agreement between the RKG and the PRG was signed on 25 September for the trade of non-military goods. This agreement and Sihanouk's move to dissolve the International Control Commission in late 1969 were seen as attempts by Sihanouk to maintain good relations with Communist China, North Vietnam and the PRG.
15. With the departure of Sihanouk for France in early January 1970, a new phase in Cambodian/Communist relations opened under acting Prime Minister Sirik Matak, who tried to pressure the Vietnamese Communists into reducing their military presence in Cambodia. A mid-January meeting between Matak and Tran Buu Kiem, a high-ranking PRG official, demonstrated the ill will between the Lon Nol- Sirik Matak government and the Vietnamese Communists. Matak reportedly accused the Communists of killing men and women

of his country at a time when Cambodia was aiding the Viet Cong and threatened to take military action against them. Kiem allegedly responded with a warning that any such action would result in hard fighting. Matak then reportedly ordered FARK to destroy Communist installations, including medical facilities in Svay Rieng Province. The Cambodian Army also continued its campaign in the northeast to try to regain some control over Communist and insurgent-held territory. ▓▓▓▓▓▓▓
▓▓▓▓▓▓▓▓▓▓▓▓▓▓▓▓▓▓▓▓▓▓▓▓▓
▓▓▓▓▓▓▓▓▓▓▓▓▓▓▓▓▓▓▓▓▓▓▓▓▓
▓▓▓▓▓▓▓ Prime Minister Lon Nol was concerned with Viet Cong support of local insurgents. An article attributed to him, which appeared in an official journal in March, gave an unusually detailed account of Cambodian guerrilla groups based in Viet Cong-controlled areas of South Vietnam that had been crossing into three southeastern border provinces to spread anti-government propaganda. The article alleged that these groups, totaling about 450 men, were led by long-time members of the Cambodian Communist Party, and

apparently planned to extend their operations westward.

16. The Cambodian government placed a 15 March deadline for VC/NVA forces to leave Cambodia. Pressure continued to build when, on 11 March, the North Vietnamese and PRG Embassies were sacked by thousands of students. Anti-Viet Cong activities also took place in several provinces. Prior to the 15 March deadline, a compromise was reported between the RKG and North Vietnam, and the parties agreed to meet on 16 March to discuss the settlement of problems.

17. ▓▓▓▓▓▓▓▓▓▓▓▓▓▓▓▓▓▓▓▓▓▓▓▓▓▓▓▓▓▓

    ▓▓▓▓▓▓ On 9 April, it was reported that Cambodian officials had taken measures to close the Tonle Kong River to Vietnamese Communist traffic. The waterway had been used since 1966 to transport supplies to Laos.

18. In response to these moves by the RKG, the Vietnamese Communists began to move

militarily in Cambodia. On 17 April, they attacked government outposts all along the frontier. The most sustained actions occurred in Svay Rieng Province, while Communist military activity rose substantially elsewhere along the border. Phnom Penh then substantially reinforced the military districts east of the capital. Later in April, the Communists continued their intensified military pressure, the heaviest action being concentrated in Kandal and Takeo Provinces, south of Phnom Penh. The situation was unstable and was rapidly deteriorating in the countryside. The Communists continued their attacks against government positions and key lines of communication in widely separated areas of eastern Cambodia. Phnom Penh was taking an increasingly tough line with the Communists.

19. The Communists at this time also began to devote considerable attention to forming a "Cambodian Liberation Movement" and to expand "liberated areas" in Cambodia. With the fall of Stung Treng on 15 may, the Communists acquired control over the Se Kong-Mekong waterway from the Laos border to the

outskirts of Kompong Cham. Lon Nol's government position deteriorated in the countryside in the face of increasing military operations. According to the late August MACV assessment, a study of enemy activities in Cambodia suggested that Hanoi probably established five near-term objectives in Cambodia: (a) to re-establish a secure LOC system in Cambodia, (b) to achieve freedom of movement in eastern Cambodia, (c) to organize a guerilla base to develop a Khmer insurgency movement, (d) to isolate Phnom Penh and other population centers, and (e) to consolidate control over the Tonle Sap area. To achieve these objectives, the enemy deployed an estimated 53,000 men in or near Cambodia.

20. The Communists continued to harass the provincial capitals and Cambodian LOCs from September to the end of the year. MACV estimated from observing the pattern of Communist activity that the enemy probably was using a protracted war strateGy. This strategy was probably designed to

    (a) cripple and demoralize FARK forces,

(b) provide a screen for logistical activities,
(c) create insecurity in the countryside,
(d) isolate the populace from the government,
(e) develop an infrastructure, and
(f) wreck the national economy.

D. **LOGISTICS**.

1. From the early 1960s into 1970, the MACV assessments concerning the Vietnamese Communist logistical system, which supported Communist forces in South Vietnam were consistent; they indicated unerringly that enemy forces in northern South Vietnam received their supplies across the DMZ and through Laos while those in southern South Vietnam received the bulk of their supplies from Cambodia once **MARKET TIME** Operations constituted an effective barrier to direct seas deliveries by the North Vietnamese by mid-1966.
2. Significantly, in August 1964, an official Vietnamese service intelligence study discussed Viet Cong supply routes. It was believed that the evidence in

the study was convincing enough to regard the report as probably true. The study in part stated, "The GVN realized the importance of the supply routes from Cambodia in relations to the increased fighting capacity of the VC in South Vietnam. At present, because of the heightened development of their forces, VC demands for equipment have increased many times over, and one of their principal concerns, therefore, is logistical operations from outside Vietnam. These operations can be implemented in three ways: (a) by sea routes from North Vietnam, (b) by land routes from Laos, and (c) by land and sea routes from Cambodia. Because of the increased patrol operations by the GVN Navy and because VC transports can be uncovered very easily, the VC have used the sea routes only with the greatest of care. Land routes through Laos, with their many mountains and other obstacles, have not been very conducive to large-scale supply movements, and have, as a result, been used principally for infiltration of cadres of South Vietnam. Therefore, because of the inadvisability of using these routes from North Vietnam and Laos very frequently, the VC have been compelled to rely almost exclusively on the routes through Cambodia for their supplies."

3. **The Sea Route from North Vietnam**

   a. Sea infiltration of materials and selected cadres into RVN by steel-hulled trawlers, capable of carrying more than 100 tons of cargo, probably started in the late 1950s. By 1960, the North Vietnamese had established an official infiltration unit, designated Group 125, which, in time, came under the direct control of the Military High Command in Hanoi. There were numerous reports from detainees and ralliers of extensive infiltration shipments being made to RVN in 1963, 1964 and early 1965. However, after the Vung Ro Bay (Phu Yen Province, RVN) incident of 16 February 1965, which involved the destruction of a Group 125 trawler loaded with arms and ammunition, a sharp decline in trawler shipments occurred as a counter-sea-infiltration effort,

MARKET TIME, began operating in March 1965 to seal the RVN coast. With the development and refinement of Operation **MARKET TIME** in 1965 and 1966, the enemy's ability to be supplied by sea infiltration of material decreased drastically. COMUSMACV estimated that, in 1965, approximately 70 percent of the NVA/VC resupply was accomplished by maritime infiltration. In mid-1966, COMUSMACV considered that the success of **MARKET TIME** operations was such that not more than 10 percent of NVA/VC logistic support came by sea. By the end of 1966, COMUSMACV was able to state that there was no credible evidence of any significant infiltration of enemy troops or supplies by sea since November 1965. All indications were that Operation **MARKET TIME** had interdicted this avenue of resupply and continued effectively to do so up through 1970.

4. (a) A MACV assessment stated that, according to a statement by Thai Foreign Minister Thanat on 29 July, Thailand

would send in troops to Cambodia "only as a last resort." The reasons for this stand, Thanat explained, were that foreign forces in a country generate friction (he cited the "acrimonious" situation that had developed in Cambodia since the introduction of RVN forces) and that the Thai government did not want to "compromise the efforts of the Djakarta group."

(b) An article in the New York Times noted that Prince Sihanouk announced on 20 September that his exile government would gradually return to Cambodian soil and that revolutionary governments had been set up in 80 of 103 villages in Kandal and Takeo Provinces.

(c) On 8 October, President Nixon announced several new initiatives for peace in Indochina, including proposals for a "cease fire in place," an Indochina Peace Conference and negotiations for complete troop withdrawals. The GVN actively backed the President's peace proposals while the Phnom Penh government was somewhat slow in publicly doing so.

VC/NVA BASE AREAS IN CAMBODIA

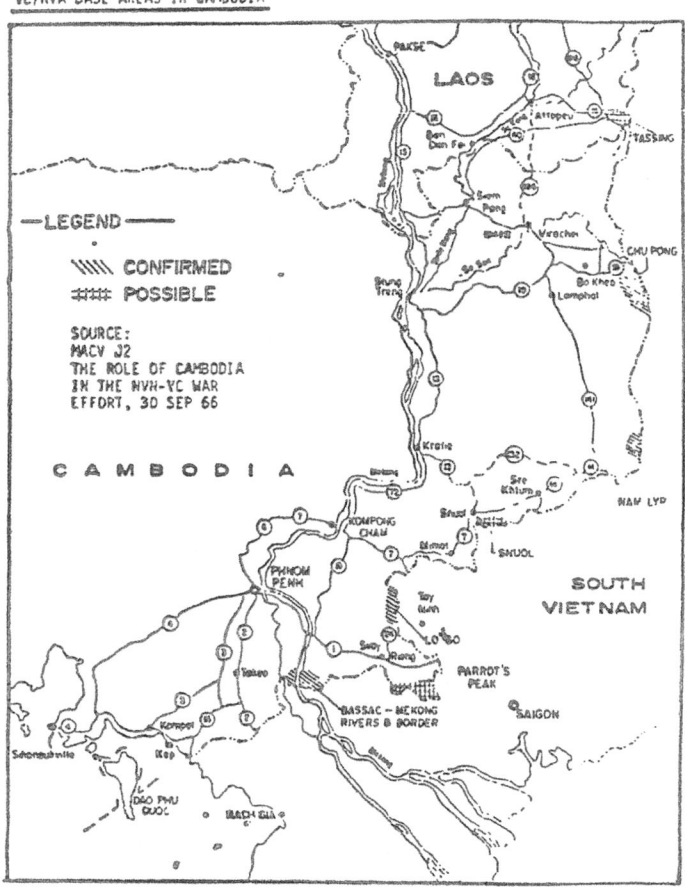

UNCLASSIFIED

THE WHITE HOUSE

WASHINGTON

PRESIDENT'S FOREIGN INTELLIGENCE ADVISORY BOARD

December 14, 1970

Dear General McChristian:

Recently, the President's Foreign Advisory Board was directed by the President to report to him on a highly complex and important development involving U. S. intelligence operations.

There was a considerable amount of urgency in rendering this report, which involved the activities of both the Washington Intelligence Community and intelligence operations conducted by U. S. Field Commanders.

In the course of our preparing its report, the Board learned that Major Frank T. McCarthy, United States Army, 483-32-2505, had been deeply involved in a related analytic problem while assigned to the J-2 division, USMACV. We also discovered that Major McCarthy had since been assigned to your command in Washington, D. C. Accordingly, I requested of the Department of the Army that Major McCarthy be permitted to assist the Board, on very short notice, in its deliberations. The Army kindly acceded to our request.

The purpose of my letter is to inform you that Major McCarthy responded to our difficult and probing questions to a degree that went far beyond our expectations. He proved to have been extremely knowledgeable in this subject matter and to have a profound knowledge of the underlining issues involved. In addition, his presentations, which he necessarily made on an extemporaneously basis, were exceptionally articulate. Our resultant report to the President was based largely upon the information he provided and upon his profound insights. Our report was well-received and appreciated by the President.

I would like to express, on behalf of the Board, my thanks to you for making available the services of this fine officer. He is a credit not only to the Army but to the Intelligence Community generally. I am sure that an officer of his outstanding qualifications will progress rapidly in the Army's Intelligence system.

With sincere best wishes,

Gerard P. Burke
Executive Secretary

Major General Joseph A. McChristian, USA
Assistant Chief of Staff for Intelligence
Department of the Army
Washington, D. C. 20310